GREATER LONDON

CEMETERIES AND CREMATORIA

AND THEIR REGISTERS

Compiled by Patricia S. Wolfston

Edited by Michael J. Gandy

Published by the Society of Genealogists
37 Harrington Gardens
London SW7 4JX

ISBN 0 901878 55 3

The London Boroughs

Table of Contents

INTRODUCTION

Locating the burial place of a 19th century resident of the Greater London area is far from an easy task, and the same applies to those who died in this century.

While many burials continued in the local churchyards of the outer London parishes, burials in the inner London churchyards or burial grounds were discontinued at various dates in the 1850s and no parish burial registers were maintained thereafter. Subsequently most burials for inner London took place further away, at non-denominational cemeteries established by either commercial companies or local government bodies.

Exceptions were those established by the Jewish community and the Roman Catholic church.

The Search

Some of the following guide lines may prove of value:

1. The deceased of certain areas of London have tended to predominate in certain cemeteries:

> residents of North London in Highgate cemetery
> residents of East London in Tower Hamlets, Victoria Park and Brookwood cemeteries
> residents of South London in Norwood and Nunhead cemeteries
> residents of West London in Kensal Green cemetery.

2. Economic status could affect the location of burial. Brookwood cemetery and others competed to undertake contracts tendered each year by several London boroughs for the burial of their poor. Although out of the Greater London Council area, Brookwood probably buried half of East London and to facilitate this Waterloo Station had a special casket-loading platform.

3. The St. Marylebone cemetery catered for the affluent middle classes of Marylebone, Highgate and Hampstead, with a high proportion also in the professional and military occupations.

4. Certain parishes bought parcels of newly established cemeteries: St. George Hanover Square at Hanwell, and St. Anne Soho, St. Margaret and St. John Westminster at Brookwood. At Norwood an area was reserved for the Brotherhood of the Greek community in London.

5. Foremost amongst the burial grounds devoted especially to Dissenters was Bunhill Fields and afterwards Abney Park cemetery. For those who were Quakers it would be advisable to undertake the search at the library of the Religious Society of Friends, Friends House, Euston Road, London NW1 (Tel. 01-387 3601).

6. Just as the names of the registration districts used by the General Register Office (St. Catherine's House) keep changing so do those of the cemeteries. As ownership changes, so do some of the names, although obviously not the location; hence if encountering difficulty it is wise to check for similarity of address.

Before contacting a repository

Not all registers contain indexes and in some cases where they exist they are for owners of grave plots rather than individuals interred there. Therefore it is strongly recommended that the approximate date of death be ascertained prior to making any contact with a source listed below.

Where it has been indicated that a member of the public MAY NOT undertake the search, we have been requested to provide:

the name of the deceased
the approximate date of death
the reason the information is being requested.

Since some repositories charge a fee, do not be surprised if that proves to be the case with the one you approach.

Where it has been indicated that a member of the public MAY undertake the search, it is recommended that an appointment be requested a minimum of 24 hours in advance. Many of these places are primarily engaged in the planning of present day funerals which must obviously take precedence.

For those interested in the history, architectural features, indentity of the well-known buried there, or the current physical conditions of a particular cemetery, several books may be consulted. The following provide an excellent introduction:

The London Burial Grounds : Mrs. Basil Holmes 1896
London Cemeteries : Hugh Meller 1981
Wate's Book of London Churchyards : Harvey Hackman 1981

The compiler is indebted to some known, and many more unknown, individuals all of whom kindly and patiently replied to numerous queries. Despite their co-operation there may be inaccuracies or omissions; and information concerning same will be most gratefully received.

GEOGRAPHICAL DISTRIBUTION

Tables designed to show the geographical area (N-E-S-W-) of all, except the Jewish, burial places considered to be 'local'; i.e. either owned by the governmental unit or physically located in said area if commercially sponsored.

Reading from left to right, the second column provides the name of the pre 1965 authority formerly in the London County Council, the old county of Middlesex, or the counties of Essex, Hertfordshire, Kent and Surrey. Column three indicates the 1982 name of the London Borough covering that same area under the Greater London Council. The final column indicates the 'local' cemetery or crematorium and in brackets the date of its first register.

area	pre-1965 authority	present authority	local cemeteries / crematoria
	City of London Corporation	City of London Corporation	Bunhill Fields Burial Grounds (1713) City of London cemetery Little Ilford (1856) City of London cemetery Aldersbrook Road (1856) City of London crematorium (1905)
N	Finsbury Islington (L.C.C.)	Islington	Islington cemetery High Road (1854) Islington cemetery Cockfosters Road (1960) Islington crematorium (1937)
N	Hornsey Tottenham Wood Green (Middlesex)	Haringey	Tottenham cemetery (1856)
N	Edmonton Enfield Southgate (Middlesex)	Enfield	Edmonton cemetery (1884) Edmonton & Southgate cemetery (ca 1880) Enfield crematorium (1938) Enfield lawn cemetery (1961) Hertford Road cemetery (1881) Lavender Hill cemetery (1872) Tottenham Park cemetery (1912)

area	pre-1965 authority	present authority	local cemeteries / crematoria
NE	Hackney Shoreditch Stoke Newington (L.C.C.)	Hackney	Abney Park cemetery (nonconformists; early 19th cen) St. Thomas Square cemetery (1837) Victoria Park cemetery (1853)
NE	Chingford Leyton Walthamstow (Essex)	Waltham Forest	Chingford Mount cemetery (1886) St. Patrick's cemetery (Catholic 1861) Walthamstow cemetery (1872)
NE	Part of Chigwell Ilford Wanstead & Woodford (Essex)	Redbridge	Barkingside cemeteries (1923) Barkingside Garden of Rest (1953) Buckingham Road cemetery (early 19th cen) Roding Lane cemetery (1940)
E	Bethnal Green Poplar Stepney (L.C.C.)	Tower Hamlets	Tower Hamlets cemetery (1841) (see introduction re Brookwood)
E	East Ham West Ham (Essex)	Newham	East London cemetery (1870) Manor Park cemetery (1874) West Ham cemetery, Forest Gate (1854) Woodgrange Park cemetery (ca 1909)
E	Barking* Dagenham (Essex)	Barking	Chadwell Heath cemetery (1937) Eastbrookend cemetery (1916) Rippleside cemetery (1886)
E	Hornchurch Romford (Essex)	Havering	Hornchurch cemetery (1932) Rainham cemetery, Upminster Road North (1902) Romford cemetery (1871) South Essex crematorium (1950) Upminster cemetery (1902)
SE	Bermondsey Camberwell Southwark (L.C.C.)	Southwark	Camberwell cemetery (1856) Camberwell New cemetery (1927) Honor Oak crematorium (1939) Nunhead cemetery (1840)
SE	Greenwich Woolwich* (L.C.C.)	Greenwich	Charlton cemetery (1864) Eltham cemetery (1935) Greenwich cemetery (1856) Plumstead cemetery (1890) Royal Hospital cemetery (Navy 1848) Woolwich cemeteries (1856)

*Two very small enclaves north of the Thames, formerly in Woolwich, and a very small area formerly in Barking but west of Barking Creek now are in Newham.

area	pre-1965 authority	present authority	local cemeteries / crematoria
SE	Deptford Lewisham (L.C.C.)	Lewisham	Brockley Ladywell cemetery (1858) Grove Park cemetery (1935) Hither Green cemetery (1873) Ladywell cemetery (1858) Lewisham crematorium (1956)
SE	Bexley Crayford Part of Chisle-hurst & Sidcup Erith (Kent)	Bexley	Bexleyheath cemetery (1894) Erith cemetery (1894) Sidcup cemetery (1894)
SE	Beckenham Bromley Part of Chisle-hurst & Sidcup Orpington Penge* (Kent)	Bromley	Beckenham Cemetery (1876) Beckenham crematorium (1956) Biggin Hill cemetery (1937) Bromley Hill cemetery (1907) Chislehurst cemetery (1912) London Road cemetery (1877) Plaistow cemetery (1895) St. Luke's cemetery (1894) St. Mary Cray cemetery (1884)
S	Croydon Coulsdon & Purley (Surrey)	Croydon	Croydon cemetery (1897) Greenlawn Memorial Park (1940) Queen's Road cemetery (1861)
SW	Lambeth Part of Wandsworth (i.e. Clapham & Streatham areas) (L.C.C.)	Lambeth	Lambeth cemetery (1854) Streatham cemetery (1893) West Norwood cemetery (1837) West Norwood crematorium (1915)
SW	Battersea Wandsworth (except for Clapham & Streatham areas) (L.C.C.)	Wandsworth	Battersea St. Mary's cemetery (1860) Putney Lower Common cemetery (1855) Putney Vale cemetery (1891) Wandsworth cemetery (1878)
SW	Merton & Mordon Mitcham Wimbledon (Surrey)	Merton	Merton & Sutton Joint cemetery (1947) Mitcham cemetery, Church Road (1883) Mitcham cemetery, London Road (1929) South London crematorium (1936) Streatham Park cemetery (1911) Streatham Park crematorium (1937) Wimbledon cemetery (1876)

*Prior to 1965 Penge was administratively part of Bromley, Kent. It was formerly in Surrey.

area	pre-1965 authority	present authority	local cemeteries / crematoria
SW	Beddington & Wallington Carshalton Sutton & Cheam (Surrey)	Sutton	Cuddington cemetery (1902) Sutton cemetery (1889)
SW	Barnes Richmond-upon-Thames (Surrey) Twickenham (Middlesex)	Richmond-upon-Thames	East Sheen cemetery (1906) Hampton cemetery (1859) Mortlake crematorium (1939)[3] Old Mortlake cemetery (1887) Richmond cemetery (1894) South West Middlesex crematorium (1954)[2] Teddington cemetery (1879) Twickenham cemetery (1867)
SW	Kingston-upon-Thames Malden & Coombe Surbiton	Kingston-upon-Thames	Kingston cemetery Kingston crematorium Surbiton cemetery
W	Fulham Hammersmith (L.C.C.)	Hammersmith & Fulham	Brompton cemetery (1840) Fulham cemetery (1865) Hammersmith cemetery (1869) Hammersmith New cemetery (1926) Part of Kensal Green cemetery (1833) Mortlake cemetery (1926) Mortlake crematorium (1939)[3] North Sheen cemetery (1905) St. Mary's cemetery (Catholic ca 1850) West London crematorium (1938)
W	St. Marylebone Paddington City of Westminster (L.C.C.)	City of Westminster	Paddington cemetery, Milespit Hill (1937) Paddington cemetery, Willesden Lane (1855) St. Marylebone cemetery (1855) St. Marylebone crematoria (1938) Westminster cemetery (1854) (see introduction re certain parishes)
W	Chelsea Kensington (L.C.C.)	Royal Borough of Kensington & Chelsea	Gunnersbury cemetery (1929) Part of Kensal Green cemetery (1833) Kensington Hanwell cemetery (1855) Royal Hospital Chelsea Burial Ground (Army 1692)

[2] South West Middlesex Crematorium is jointly owned by Ealing, Hillingdon, Hounslow, Richmond and Spelthorne.

[3] Mortlake Crematorium is jointly owned by Ealing, Hammsersmith, Hounslow and Richmond.

area	pre-1965 authority	present authority	local cemeteries / crematoria
W	Acton Ealing Southall (Middlesex)	Ealing	Acton cemetery (1895) Greenford Park cemetery (1901) Havelock Road cemetery (1883) Hortus Road cemetery (1944) Mortlake crematorium (1939)[3] South Ealing cemetery (1861) South West Middlesex crematorium (1954)[2]
W	Brentford & Chiswick Heston & Isleworth (Middlesex)	Hounslow	Bedfont cemetery (1941) Borough cemetery (1942) Chiswick New cemetery (1933) Chiswick Old cemetery (1883) Feltham cemetery (1886) Hatton cemetery (1974) Hounslow cemetery (1869) Isleworth cemetery (1880) Mortlake crematorium (1939)[3] New Brentford cemetery (1903) South West Middlesex crematorium (1954)[2]
W	Hayes & Harlington Ruislip – Northwood Uxbridge Yiewsley & West Drayton (Middlesex)	Hillingdon	Breakspear crematorium (1958)[1] Cherry Lane cemetery (1937) Harlington Burial Ground (1871) Harmondsworth cemetery (1905) Hillingdon & Uxbridge cemetery (1856) Northwood cemetery (1915) South West Middlesex crematorium (1954)[2] Victoria Lane Burial Ground (1871) West Drayton cemetery (1939)
NW	Hampstead Holborn St. Pancras (L.C.C.)	Camden	Hampstead cemetery (1876) Highgate cemetery (1839) St. Pancras cemetery (1854)
NW	Wembley Willesden (Middlesex)	Brent	Alperton cemetery (1917) Carpenders Park cemetery (1954) Willesden New cemetery (1893)

[1] Breakspear Crematorium is jointly owned by Harrow and Hillingdon.
[2] South West Middlesex Crematorium is jointly owned by Ealing, Hillingdon, Hounslow, Richmond and Spelthorne.
[3] Mortlake Crematorium is jointly owned by Ealing, Hammersmith, Hounslow and Richmond.

area	pre-1965 authority	present authority	local cemeteries / crematoria
NW	Harrow Middlesex	Harrow	Breakspear crematorium (1958)[1] Eastcote Lane cemetery (1922) Harrow cemetery (1888) Harrow Weald cemetery (1937) Pinner Lane cemetery (1860s) Pinner New cemetery (1933) Roxeth Hill Burial Ground (1922) Wealdstone cemetery (1902)
NW	Barnet East Barnet (Herts.) Finchley Friern Barnet Hendon (Middlesex)	Barnet	Golders Green crematorium (1903) Great Northern London cemetery (1861) Hendon cemetery (1899) Hendon crematorium (1922)

MIDDLESEX EXCLUDED FROM G.L.C.

	Potters Bar	Hertsmere, Hertfordshire	Allum Lane Lawn cemetery
	Staines Sunbury-on-Thames	Spelthorne, Surrey	Ashford cemetery (1910) South West Middlesex crematorium (1954) Staines cemetery (1913) Stanwell Burial Ground (1900) Sunbury cemetery (1900)

OUTSIDE G.L.C. AREA

Brookwood cemetery, Woking, Surrey (ca 1854)

[1] Breakspear Crematorium is jointly owned by Harrow and Hillingdon.

NON-DENOMINATIONAL CEMETERIES AND CREMATORIA

ALPHABETICAL LISTING

Name & address of cemetery	Location of Burial Registers	Date Registers Begin	Can you yourself search
Abney Park Cemetery Stoke Newington High Street London N16	at the cemetery Tel. 01-249 8928	Early 19th Cen.	Yes
Acton Cemetery Park Royal Road London NW10	Cemetery Admin. Office London Borough of Ealing Parks Division Central Sports Ground Horsenden Lane South Greenford Middlesex Tel. 01-998 5441	1895	Yes
All Saints Cemetery Clifford Road Alperton Wembley Middlesex	Cemeteries Office London Borough of Brent Clifford Road, Alperton Wembley, Middlesex HA0 1AF Tel. 01-902 2385	1917	
All Saints Cemetery – see Nunhead			
All Souls Cemetery – see Kensal Green			
Allum Lane Lawn Cemetery Allum Lane Borehamwood, Herts.	Hertsmere Borough Council Civic Offices, Elstree Way Borehamwood, Herts. Tel. 01-207 2277	1962	?
Alperton Cemetery Clifford Road, Alperton Wembley, Middlesex	Cemeteries Office London Borough of Brent Clifford Road, Alperton Wembley, Middlesex HA0 1AF Tel. 01-902 2385	1917	Yes
Ashford Cemetery Long Lane Ashford, Middlesex	Spelthorne Borough Council Fordbridge Park 325 Kingston Road Ashford, Middlesex Tel. Ashford (69) 55641	1910	No
Barkingside Cemeteries Longwood Gardens Ilford, Essex	Longwood Gardens Tel. 01-550 5267	1923	No

Name & address of cemetery	Location of Burial Registers	Date Registers Begin	Can you yourself search
Barkingside Garden of Rest Longwood Gardens Ilford, Essex	Longwood Gardens Tel.01-550 5267	1953	No
Battersea St. Mary's Cemetery Bolingbroke Grove London SW11	Wandsworth Cemetery Magdalen Road Wandsworth SW18 3NP	1860	No
Beckenham Cemetery Elmers End Road Beckenham, Kent	at the cemetery Tel. 01-650 0322	1876	Yes
Beckenham Crematorium Elmers End Road Beckenham, Kent	at Beckenham Cemetery Tel. 01-650 0322	1956	Yes
Bedfont Cemetery Bedfont Road Feltham, Middlesex	Civic Centre London Borough of Hounslow Lampton Road Hounslow TW3 4DN Tel. cemetery section 01-570 7728	1941	Yes
Bexleyheath Cemetery Banks Lane Broadway Bexleyheath, Kent	London Borough of Bexley Cemetery Department 6 Gravel Hill Bexleyheath, Kent DA6 7PN Tel. 01-303 7777 x 2316	1894	No
Biggin Hill Cemetery Kingsmead Avenue Biggin Hill, Kent	Cemeteries Section Department of Recreation Bromley Civic Centre Rochester Avenue Bromley, Kent BR1 3UH	1937	No
Borough Cemetery Powder Mill Lane Twickenham, Middlesex	as for Bedfont	1942	Yes
Breakspear Crematorium Breakspear Road Ruislip, Middlesex	at the crematorium Tel. 01-713 2843	1958	No
Brockley Ladywell Cemetery Brockley Road London SE13	Ladywell Cemetery Ladywell Road, London SE13	1858	Yes
Bromley Hill Cemetery Bromley Hill Bromley, Kent	Cemeteries Section Department of Recreation Bromley Civic Centre Rochester Avenue Bromley, Kent BR1 3UH Tel. 01-464 3333 x 3256	1907	No

Name & address of cemetery	Location of Burial Registers	Date Registers Begin	Can you yourself search
Brompton Cemetery Fulham Road London SW10	at the cemetery Tel. 01-352 1201	1840	No
Brookwood Cemetery Cemetery Pales Brookwood, Woking Surrey GU24 0BL	at the cemetery Tel. (04867)2222	ca 1854	No
Buckingham Road Cemetery High Street Ilford, Essex	at the cemetery Tel. 01-478 8510	early 19th Cen.	No
Bunhill Fields Burial Ground City Road London EC1	1713-1854 indexes at Public Record Office Chancery Lane (RG4/4652-4657)		Yes
	1789-1854 interment order books and transcripts of inscriptions (Ms 1092): 18 vols. at Guildhall library		Yes
	1823-1854 index to books of interment at Society of Genealogists library		Yes
Camberwell Old Cemetery Forest Hill Road London SE22	Camberwell New Cemetery Office Brenchley Gardens London SE23 Tel. 01-639 3121	1856	Yes
Camberwell New Cemetery Brenchley Gardens London SE23	at the cemetery Tel. 01-639 3121	1927	Yes
Carpenders Park Cemetery Oxhey Lane Watford, Herts.	Cemeteries Office London Borough of Brent Clifford Road, Alperton Wembley, Middlesex HA0 1AF	1954	Yes
Chadwell Heath Cemetery Whalebone Lane North Dagenham, Essex	Rippleside Cemetery Ripple Road, Barking, Essex Tel. 01-594 2656	1937	Yes
Charlton Cemetery Cemetery Lane London SE7	at the cemetery and also at Admin. Offices, Room 30 Town Hall, Woolwich, London SE18	1864	Yes
Cherry Lane Cemetery Shepiston Lane Harlington, Middlesex	Cemeteries Department Civic Centre, High Street Uxbridge, Middlesex Tel. Uxbridge (0895)50111 x 2382/3	1937	Yes

Name & address of cemetery	Location of Burial Registers	Date Registers Begin	Can you yourself search
Chingford Mount Cemetery Old Church Road London E4	at the cemetery Tel. 01-524 5030	1886	No
Chislehurst Cemetery Beaverwood Road Chislehurst, Kent	Cemeteries section Department of Recreation Bromley Civic Centre Rochester Avenue Bromley, Kent BR1 3UH Tel. 01-464 3333 x 3256	1912	No
Chiswick New Cemetery Staveley Road London W4	part at cemetery; part at Civic Centre London Borough of Hounslow Lampton Road, Hounslow TW3 4DN Tel. cemetery section 01-570 7728	1933	Yes
Chiswick Old Cemetery Corney Road London W4	part at Chiswick New Cemetery Staveley Road, London W4 part at Civic Centre London Borough of Hounslow Lampton Road, Hounslow TW3 4DN Tel. cemetery section 01-570 7728	1888	Yes
City of London and Tower Hamlets – see Tower Hamlets Cemetery			
City of London Cemetery Little Ilford Essex	Guildhall library (Ms 10445); 61 vols	1856- 1915	Yes
City of London Cemetery Aldersbrook Road Manor Park, London E12	at the cemetery Tel. 01-530 2151	1856	Yes
City of London Crematorium Aldersbrook Road Manor Park, London E12	City of London Cemetery Aldersbrook Road Manor Park, London E12	1905	Yes
City of Westminster – see Westminster Cemetery			
Croydon Cemetery Mitcham Road Croydon, Surrey	at the cemetery Tel. 01-684 3877	1897	Yes
Crystal Palace District Cemetery – see Beckenham Cemetery			
Cuddington Cemetery Lindsey Road, Worcester Park Sutton, Surrey	Sutton Cemetery Alcorn Close, Sutton, Surrey	1902	No
Deptford Cemetery – see Brockley Cemetery & Ladywell Cemetery			

Name & address of cemetery	Location of Burial Registers	Date Registers Begin	Can you yourself search
Ealing & Old Brentford Cemetery – see South Ealing Cemetery			
Eastbrook Cemetery The Chase, off Dagenham Road Dagenham, Essex	Rippleside Cemetery Ripple Road, Barking, Essex Tel. 01-594 2656	1916	Yes
Eastcote Lane Cemetery Eastcote Lane South Harrow, Middlesex	New Pinner Cemetery Pinner Road, Harrow, Middlesex Tel. 01-866 1860	1922	No
East Finchley Cemetery – see Islington, St. Pancras & St. Marylebone Cemeteries			
East London Cemetery Grange Road Plaistow, London E13	at the cemetery	1874	No
East Sheen Cemetery Sheen Road Richmond, Surrey	Richmond Cemetery Office Grove Road, Richmond, Surrey	1906	No
Edmonton Cemetery Church Street London N9	part at cemetery, part at Cemetery Department The Civic Centre, Enfield, Middlesex Tel. 01-366 6565	1884	Yes
Edmonton & Southgate Cemetery Waterfall Road London N11	Cemetery Department The Civic Centre, Enfield, Middlesex Tel. 01-366 6565	ca 1880	Yes
Elmers End Road Cemetery – see Beckenham Cemetery			
Eltham Cemetery Rochester Way London SE9	at cemetery and also at Admin. Offices, Room 30 Town Hall, Woolwich, London SE18	1935	Yes
Enfield Cemetery – see Hertford Road Cemetery & Lavender Hill Cemetery			
Enfield Crematorium Great Cambridge Road Enfield, Middlesex EN1 4DS	at the crematorium Tel. 01-363 8324	1938	Yes
Enfield Lawn Cemetery Enfield Crematorium Great Cambridge Road Enfield, Middlesex EN1 4DS	Enfield Crematorium Tel. 01-363 8324	1961	Yes
Erith Cemetery Brook Street Erith, Kent	London Borough of Bexley Cemetery Department 6 Gravel Hill, Bexleyheath Kent DA6 7PN	1894	No

Name & address of cemetery	Location of Burial Registers	Date Registers Begin	Can you yourself search
Feltham Cemetery Sunbury Road Feltham, Middlesex	as for Bedfont	1886	Yes
Fulham Cemetery Fulham Palace Road London SW6	Parks & Cemetery Department London Borough of Hammersmith & Fulham, Pryors Bank Bishop's Park, London SW6 Tel. 01-736 7181	1865	No
Golders Green Crematorium Hoop Lane London NW11	at the crematorium Tel. 01-455 2374	1903	No
Great Northern London Cemetery Brunswick Park Road London N11	at the cemetery Tel. 01-368 2677	1861	No
Greenford Park Cemetery Windmill Lane Greenford, Middlesex	Cemetery Admin. Office London Borough of Ealing Parks Division Ealing Central Sports Ground Horsenden Lane South Greenford, Middlesex Tel. 01-998 5441	1901	Yes
Greenlawn Memorial Park Chelsham Warlingham, Surrey	Croydon Cemetery Mitcham Road, Croydon, Surrey Tel. 01-684 3877	1940	Yes
Greenwich Cemetery Well Hall Road London SE9	at the cemetery and also at Admin. Offices, Room 30 Town Hall Woolwich, London SE18	1856	Yes
Greenwich Hospital Burial Ground – see Royal Hospital Cemetery			
Grove Park Cemetery Marvels Lane Lee, London SE12	at the cemetery	1935	Yes
Gunnersbury Cemetery Gunnersbury Avenue London W4	at the cemetery Tel. 01-992 2924	1929	No
Hammersmith Cemetery Margravine Road London W6	Parks & Cemetery Department London Borough of Hammersmith & Fulham, Pryor's Bank Bishop's Park, London SW6 Tel. 01-736 7181	1869	No

Name & address of cemetery	Location of Burial Registers	Date Registers Begin	Can you yourself search
Hammersmith New Cemetery Clifford Avenue London SW14	Parks & Cemetery Department London Borough of Hammersmith & Fulham, Pryors Bank Bishops Park, London SW6 Tel. 01-736 7181	1926	No
Hampstead Cemetery Fortune Green Road London NW6	at the cemetery Tel. 01-435 6142	1876	No
Hampton Cemetery Broad Lane, Hampton, Middlesex	Twickenham Cemetery Hospital Bridge Road Twickenham, Middlesex	1859	No
Hanwell Cemetery – see Kensington Hanwell, also Westminster			
Harlington Burial Ground St. Peter's Way Harlington, Middlesex	Cemetery Department, Civic Centre High Street, Uxbridge, Middlesex Tel. Uxbridge (0895)50111 x 2382/3	1871	Yes
Harmondsworth Cemetery High Street Harmondsworth Village Middlesex	Cemetery Department, Civic Centre High Street, Uxbridge, Middlesex Tel. Uxbridge (0895)50111 x 2382/3	1905	Yes
Harrow Cemetery Pinner Road Harrow, Middlesex	Harrow Weald Cemetery Clamp Hill, Stanmore, Middlesex Tel. 01-954 1561	1888	No
Harrow Weald Cemetery Clamp Hill Road Stanmore, Middlesex	at the cemetery Tel. 01-954 1561	1937	No
Hatton Cemetery Faggs Road Feltham, Middlesex	As for Bedfont	1974	Yes
Havelock Road Cemetery Southall, Middlesex	Cemetery Admin. Office London Borough of Ealing Parks Division Central Sports Ground Horsenden Lane South Greenford, Middlesex Tel. 01-998 5441	1883	Yes
Hendon Cemetery Holders Hill Road London NW4	at the cemetery	1899	Yes
Hendon Crematorium Holders Hill Road London NW4	Hendon Cemetery Holders Hill Road London NW4	1922	Yes

Name & address of cemetery	Location of Burial Registers	Date Registers Begin	Can you yourself search
Hertford Road Cemetery Hertford Road Enfield, Middlesex	Cemetery Department, Civic Centre Enfield, Middlesex Tel. 01-366 6565	1881	Yes
Highgate Cemetery Swains Lane London N6	Holborn local history library Theobalds Road, London WC1	1839	Yes
Hillingdon & Uxbridge Cemetery Hillingdon Hill Hillingdon, Middlesex	Cemetery Department, Civic Centre High Street, Uxbridge, Middlesex Tel. Uxbridge (0895)50111 x 2382/3	1856	Yes
Hither Green Cemetery Verdant Lane Catford, London SE6	Lewisham Crematorium Verdant Lane, Catford London SE6	1873	Yes
Honor Oak Crematorium London SE23	at the crematorium Tel. 01-639 7499	1939	Yes
Hornchurch Cemetery Upminster Road Hornchurch, Essex	Havering Council Offices Mercury House, Mercury Gardens Romford, Essex RM1 3DX Tel. Romford (70)66999	1932	Yes
Hortus Road Cemetery Southall, Middlesex	Cemetery Admin. Office London Borough of Ealing Parks Division Central Sports Ground Horsenden Lane South Greenford, Middlesex Tel. 01-998 5441	1944	Yes
Hounslow Cemetery Hanworth Road Hounslow, Middlesex	as for Bedfont	1869	Yes
Isleworth Cemetery Park Road Isleworth Middlesex	Civic Centre London Borough of Hounslow Lampton Road Hounslow TWN3 4DN Tel. cemetery section 01-570 7728	1880	Yes
Islington Cemetery High Road, East Finchley London N2	at the cemetery Tel. 01-883 1230	1854	Yes
Islington Cemetery (Trent Park) Cockfosters Road Barnet, Hertfordshire	Islington Cemetery High Road, East Finchley London N2 Tel. 01-883 1230	1960	Yes

Name & address of cemetery	Location of Burial Registers	Date Registers Begin	Can you yourself search
Islington Crematorium High Road London N2	Islington Cemetery High Road, East Finchley London N2 Tel. 01-883 1230	1937	Yes
Kensal Green (All Souls) Cemetery Harrow Road London W10	at the cemetery Tel. 01-969 0152	1833	No
Kensington Hanwell Cemetery Broadway London W7	Gunnersbury Cemetery Gunnersbury Avenue, London W4 Tel. 01-992 2924	1855	No
Kingston Cemetery Bonner Hill Road Kingston-upon-Thames Surrey	at the cemetery Tel. 01-546 4462	1855	No
Kingston Crematorium Bonner Hill Road Kingston-upon-Thames Surrey	at the crematorium Tel. 01-546 4462	1952	No
Ladywell Cemetery Ladywell Road London SE13	at the cemetery Tel. 01-690 0995	1858	Yes
Lambeth Cemetery Blackshaw Road London SW17	except for the first register (in Lambeth archives) at the cemetery Tel. 01-672 1390	1854	No
Lavender Hill Cemetery Cedar Road Enfield, Middlesex	The Cemetery Department The Civic Centre Enfield, Middlesex Tel. 01-366 6565	1872	Yes
Lee Cemetery – see Hither Green Cemetery			
Lewisham Cemetery – see Brockley Cemetery and Ladywell Cemetery			
Lewisham Crematorium Verdant Lane London SE6	at the crematorium Tel. 01-698 4955	1956	Yes
London Road Cemetery Warner Road Bromley, Kent	Cemeteries Section Department of Recreation Bromley Civic Centre Rochester Avenue Bromley, Kent BR1 3UH Tel. 01-464 3333 x 3256	1877	No

Name & address of cemetery	Location of Burial Registers	Date Registers Begin	Can you yourself search
Manor Park Cemetery Sebert Road, Manor Park London E7	at the cemetery – must write	1874	No
Manor Park Crematorium Sebert Road, Manor Park London E7	at Manor Park Cemetery – must write	1955	No
Margravine Road Cemetery – see Hammersmith Cemetery			
Merton and Sutton joint Cemetery Lower Morden Drive Morden, Surrey	Cemetery Office Morden Park House London Road, Morden, Surrey Tel. 01-640 6721	1947	No
Mill Hill Cemetery – see Paddington Cemetery			
Mitcham Cemetery Church Road Mitcham, Surrey	Cemetery Office Morden Park House London Road, Morden, Surrey Tel. 01-640 6721	1883	
Mitcham Cemetery London Road Mitcham, Surrey	Cemetery Office Morden Park House London Road, Morden, Surrey Tel. 01-640 6721	1929	No
Mortlake Cemetery Clifford Avenue London SW14	Parks and Cemetery Department London Borough of Hammersmith & Fulham, Pryors Bank Bishops Park, London SW6 Tel. 01-736 7181	1926	No
Mortlake Crematorium Clifford Avenue London SW14	at the crematorium Tel. 01-876 4358	1939	No
New Brentford Cemetery Sutton Lane Hounslow, Middlesex	as for Bedfont	1903	Yes
New Southgate Cemetery & Crematorium – see Great Northern London Cemetery			
North Sheen Cemetery Lower Richmond Road London SW14	Parks & Cemetery Department London Borough of Hammersmith & Fulham, Pryors Bank Bishops Park, London SW6 Tel. 01-736 7181	1905	

Name & address of cemetery	Location of Burial Registers	Date Registers Begin	Can you yourself search
Northwood Cemetery Chesnut Avenue Northwood, Middlesex	Cemeteries Department Civic Centre High Street Uxbridge, Middlesex Tel. Uxbridge (0895)50111 x 2382/3	1915	Yes
Norwood Cemetery – see West Norwood Cemetery			
Nunhead Cemetery Linden Grove London SE15	Camberwell New Cemetery Brenchley Gardens London SE23 Tel. 01-639 3121	1840	Yes
Old Mortlake Cemetery South Worple Way London SW14	Richmond Cemetery Office Grove Road Richmond-upon-Thames Surrey Tel. 01-940 0365	1887	No
Paddington Cemetery Mill Hill, Milespit Hill London NW7	at the cemetery Tel. 01-959 2720	1937	Yes
Paddington Cemetery Willesden Lane London NW6	Cemetery Office, Milespit Hill London NW7	1855	Yes
Paines Lane Cemetery Paines Lane Pinner, Middlesex	part at Pinner New Cemetery Pinner Road, Harrow, Middlesex part at Central Library Civic Centre, Harrow, Middlesex	1860s	Yes
Pinner New Cemetery Pinner Road Pinner, Middlesex	at the cemetery Tel. 01-866 1860	1933	No
Pinner Road Cemetery – see Harrow Cemetery			
Plaistow Cemetery Burnt Ash Lane Bromley, Kent	Cemeteries Section Department of Recreation Bromley Civic Centre Rochester Avenue Bromley, Kent BR1 3UH Tel. 01-464 3333 x 3256	1895	No
Plumstead Cemetery Wickham Lane London SE2	at the cemetery and also at Admin. Offices, Room 30 Town Hall, Woolwich London SE18	1890	Yes

Name & address of cemetery	Location of Burial Registers	Date Registers Begin	Can you yourself search
Putney Lower Common Cemetery Mill Hill Road London SW13	Putney Vale Cemetery Kingston Road, Putney London SW15 3SB Tel. 01-788 2113	1855	No
Putney Vale Cemetery Kingston Road Putney, London SW15	at the cemetery Tel. 01-788 2113	1891	No
Queen's Road Cemetery Queen's Road Croydon, Surrey	Croyden Cemetery Mitcham Road Croydon, Surrey Tel. 01-684 3877	1861	Yes
Rainham Cemetery Upminster Road North Rainham, Essex	Havering Council Offices Mercury House, Mercury Gardens Romford, Essex RM1 3DX Tel. Romford (70) 66999	1902	Yes
Richmond Cemetery Grove Road Richmond-upon-Thames Surrey	at the cemetery Tel. 01-940 0365	1894	No
Rippleside Cemetery Ripple Road Barking, Essex	at the cemetery Tel. 01-594 2656	1886	Yes
Roding Lane Cemetery Roding Lane North South Woodford London E18	Barkingside Cemetery Longwood Gardens Ilford, Essex Tel. 01-550 5267	1940	No
Romford Cemetery Dagenham Road Romford, Essex	Havering Council Offices Mercury House, Mercury Gardens Romford, Essex RM1 3DX Tel. Romford (70) 66999	1871	Yes
Roxeth Hill Burial Ground Roxeth Hill Harrow, Midddlesex	Pinner New Cemetery Pinner Road, Harrow, Middlesex Tel. 01-866 1860	1922	No
Royal Hospital Cemetery (Navy) East Greenwich	Public Record Office Chancery Lane (RG8/16-18)	1848- 1864	Yes
Royal Hospital Chelsea Burial Ground (Army) Royal Hospital Road London SW3	Public Record Office Chancery Lane (RG4)	1692- 1856	

Name & address of cemetery	Location of Burial Registers	Date Registers Begin	Can you yourself search
St. Luke's Cemetery Magpie Hall Lane Bromley, Kent	Cemeteries Section Department of Recreation Bromley Civic Centre Rochester Avenue Bromley, Kent BR1 3HU Tel. 01-464 3333 x 3256	1894	No
St. Mary Cray Cemetery Star Lane St. Mary Cray Orpington, Kent	Cemeteries Section Department of Recreation Bromley Civic Centre Rochester Avenue Bromley, Kent BR1 3HU Tel. 01-464 3333 x 3256	1884	No
St. Marylebone Cemetery East End Road, East Finchley London N2	at the cemetery	1855	Yes
St. Marylebone Crematorium East End Road, East Finchley London N2	at the crematorium Tel. 01-346 8973	1938	No
St. Pancras Cemetery High Road, East Finchley London N2	at the cemetery Tel. 01-883 1231	1854	Yes
St. Thomas Square Cemetery Mare Street Hackney	Public Record Office Chancery Lane (RG8/41)	1837- 1876	Yes
Sidcup Cemetery Sidcup, Kent	London Borough of Bexley Cemetery Office, 6 Gravel Lane Bexleyheath, Kent DA6 7PN	1894	No
South Ealing Cemetery South Ealing Road London W5	Cemetery Admin. Office London Borough of Ealing, Parks Div. Ealing Central Sports Ground Horsenden Lane South Greenford, Middlesex Tel. 01-998 5441	1861	Yes
South Essex Crematorium Ockenden Road Corbets Tey Upminster, Essex RM14 2UY	at the crematorium Tel. Upminster (86) 26167	1950	Yes
South London Crematorium Rowan Road London SW16	Streatham Park Cemetery Rowan Road, London SW16 Tel. 01-764 2255	1936	No

South Metropolitan Cemetery – see West Norwood Cemetery

Name & address of cemetery	Location of Burial Registers	Date Registers Begin	Can you yourself search
South West Middlesex Crematorium, Hounslow Road Hanworth, Feltham Middlesex	at the crematorium Tel. 01-894 9001	1954	No
Staines Cemetery London Road Staines, Middlesex	Spelthorne Borough Council Fordbridge Park 325 Kingston Road Ashford, Middlesex Tel. Ashford (69) 55641	1913	No
Stanwell Burial Ground Town Lane, Stanwell Staines, Middlesex	Spelthorne Borough Council Fordbridge Park 325 Kingston Road Ashford, Middlesex Tel. Ashford (69) 55641	1900	No
Streatham Cemetery Garrat Lane London SW17	at the cemetery Tel. 01-672 1386	1893	No
Streatham Park Cemetery Rowan Road London SW16	at the cemetery Tel. 01-764 2255	1911	No
Streatham Park Crematorium Rowan Road London SW16	Streatham Park Cemetery Rowan Road, London SW16 Tel. 01-764 2255	1937	No
Sunbury Cemetery Greenway Sunbury, Middlesex	Spelthorne Borough Council Fordbridge Park 325 Kingston Road Ashford, Middlesex Tel. Ashford (69) 55641	1900	No
Surbiton Cemetery Lower Marsh Lane Kingston-upon-Thames Surrey	Kingston Cemetery Bonner Hill Road Kingston-upon-Thames Surrey Tel. 01-546 4462	1915	No
Sutton Cemetery Alcorn Close Sutton, Surrey	at the cemetery Tel. 01-644 9437	1889	No
Teddington Cemetery Shacklegate Lane Teddington TW11 8SF	Twickenham Cemetery Hospital Bridge Road Twickenham, Middlesex Tel. 01-894 2535	1879	No

Name & address of cemetery	Location of Burial Registers	Date Registers Begin	Can you yourself search
Tottenham Cemetery White Hart Lane Tottenham, London N17	Enfield Crematorium Great Cambridge Road Enfield, Middlesex EN1 4DS Tel. 01-363 8324	1856	Yes
Tottenham Park Cemetery Montagu Road London N18	at the cemetery Tel. 01-807 1617	1912	No
Tottenham & Wood Green Crematorium – see Enfield Crematorium			
Tower Hamlets Cemetery Southern Grove, London E3	Greater London Record Office County Hall, London SE1	1841	Yes
Trent Park Cemetery – see Islington Cemetery, Cockfosters Road			
Twickenham Cemetery Hospital Bridge Road Twickenham, Middlesex	at the cemetery Tel. 01-894 2535	1867	No
Upminster Cemetery Ockenden Road Corbets Tey, Upminster Essex RM14 2UY	Havering Council Offices Mercury House, Mercury Gardens Romford, Essex RM1 3DX Tel. Romford (70) 66999	1902	Yes
Uxbridge Cemetery – see Hillingdon & Uxbridge Cemetery			
Victoria Lane Burial Ground Victoria Lane Harlington, Middlesex	The Cemetery Department Civic Centre, High Street Uxbridge, Middlesex Tel. Uxbridge (0895)50111 x 2382/3	1871	Yes
Victoria Park Cemetery Hackney London E3	Public Record Office Chancery Lane (RG8/42-51)	1853 -1876	Yes
Walthamstow Cemetery Queen's Road London E17	Chingford Mount Cemetery Old Church Road London E4 Tel. 01-524 5030	1872	No
Wandsworth Cemetery Magdalen Road London SW18 3NP	at the cemetery Tel. 01-874 8386	1878	No
Wealdstone Cemetery at rear of Byron Recreation Ground Peel Road, Wealdstone Harrow, Middlesex	Harrow Weald Cemetery Clamp Hill Stanmore, Middlesex Tel. 01-954 1561	1902	No

Name & address of cemetery	Location of Burial Registers	Date Registers Begin	Can you yourself search
West Drayton Cemetery Harmondsworth Road West Drayton, Middlesex	Cemetery Department, Civic Centre High Street, Uxbridge, Middlesex Tel. Uxbridge (0895)50111 x 2382/3	1939	Yes
West Ham Cemetery Cemetery Road Forest Gate, London E7	at the cemetery Tel. 01-534 1566	1854	Yes
West London and Westminster Cemetery – see Brompton Cemetery			
West London Crematorium Harrow Road London W10	Kensal Green Cemetery Harrow Road, London W10 Tel. 01-969 0152	1938	No
Westminster Cemetery Uxbridge Road Hanwell, London W7	at the cemetery Tel. 01-567 0913	1854	Yes
West Norwood Cemetery Norwood Road London SE27 9JU	at the cemetery Tel. 01-670 0011	1837	No
West Norwood Crematorium Norwood Road London SE27 9JU	West London Cemetery Norwood Road London SE27 9JU	1915	No
Whitton Cemetery – see Twickenham Cemetery			
Willesden Lane Cemetery – see Paddington Cemetery			
Willesden New Cemetery Franklyn Road London NW10	Cemeteries Office London Borough of Brent Clifford Road, Alperton Wembley, Middlesex HA0 1AF	1893	Yes
Wimbledon Cemetery Gap Road London SW19	Cemetery Office Morden Park House London Road, Morden, Surrey	1876	No
Woodgrange Park Cemetery Romford Road London E7	at the cemetery Tel. 01-472 3433	ca 1909	Yes
Woolwich Cemeteries King's Highway & Camdale Road London SE18	at the cemetery and also at Admin. Offices, Room 30 Town Hall, Woolwich, London SE18	1856	Yes

RESTRICTED CEMETERIES AND CREMATORIA

ARMY

Royal Hospital Chelsea Burial Ground London SW3	Public Record Office Chancery Lane (RG4)	1692 -1856	Yes

NAVY

Royal Hospital Cemetery East Greenwich	Public Record Office Chancery Lane (RG8/16-18)	1848 -1864	Yes

ROMAN CATHOLIC

Mortlake Catholic Cemetery North Worple Way London SW14	St. Mary Magdalen North Worple Way Tel. 01-876 1326	1852	Yes
St. Mary's Cemetery Kensal Green, Harrow Road London NW10	at the cemetery Tel . 01-969 1145	ca 1850	No
St. Patrick's Leytonstone Cemetery, Langthorne Road London E11	at the cemetery Tel . 01-539 2451	1861	yes
South London Crematorium Rowan Road London SW16	although separated into its own physical area there are no separated registers –see above lists for details	1936	No

JEWISH

With this group the important factor is the organization rather than the residential location of the deceased.

Adath Yisroel Cemetery Carterhatch Lane Enfield, Middlesex	Adath Yisroel Burial Society 40 Queen Elizabeth Walk London N16	1925	Yes
Adath Yisroel Cemetery Silver Street Cheshunt, Hertfordshire	as above	1964	Yes
Jewish Cemetery Montagu Road Lower Edmonton, London N18	Federation of Synagogues Burial Society 9 Greatorex Street London E1 Tel . 01-247 4471	1887	No
Rainham Federation Cemetery Upminster Road North Rainham, Essex	as above	1937	No

Name & address of cemetery	Location of Burial Registers	Date Registers Begin	Can you yourself search
Edgwarebury Cemetery Edgware, Middlesex	Spanish & Portuguese Congregation 2 Ashforth Road London W9	1970s	No
Hoop Lane Cemetery Hoop Lane, London NW11	as above	late 18th cen.	No
Sephardi Novo Cemetery Mile End Road London E1	as above	1735	No
Sephardi Velho (Old) Cemetery Mile End Road London E1	as above	1656 -1735	No
Edgwarebury Cemetery Edgware, Middlesex	Union of Liberal and Progressive Synagogues Montagu Centre 109 Whitfield Street London W1 Tel. 01-580 1663/4	1976	No
Liberal Jewish Synagogue Cemetery Pound Lane Willesden, London NW10	as above	1914	No
Alderney Road Cemetery Alderney Road London E1	United Synagogue Woburn Place Upper Woburn Place London WC1H 0EZ	1696	No
Brady Street Cemetery Brady Street, London E1	as above	1761	No
Bushey Cemetery Little Bushey Lane Bushey, Hertfordshire	as above	1947	No
East Ham Cemetery Marlow Road High Street South London E6	as above	1919	No
Lauriston Road Cemetery Lauriston Road Hackney London E9	as above	1788	No

Name & address of cemetery	Location of Burial Registers	Date Registers Begin	Can you yourself search
Plashet Cemetery High Street North London E12	as above	1896	No
Waltham Abbey Cemetery Skillet Hill (Honey Lane) Waltham Abbey, Essex	as above	1960	No
West Ham Cemetery Buckingham Road Stratford, London E15	as above	1857	No
Willesden Cemetery Beaconsfield Road London NW10	as above	1873	no
Jewish Cemetery Rowan Road Streatham Vale London SW16	West End Great Synagogue 21 Dean Street London W1	1915	No
Balls Pond Cemetery Kingsbury Road London N1	West London Synagogue 33 Seymour Place London W1H 5AP Tel. 01-723 4404	1843	Yes
Edgwarebury Cemetery Edgware, Middlesex	as above	1976	Yes
Golders Green Jewish Cemetery Hoop Lane London NW11	as above	1897	Yes
Western Cemetery Bulls Cross Ride Cheshunt, Waltham Cross Hertfordshire	Western Synagogue Crawford Place London W1 Tel. 01-723 9333	1957	No
Western Cemetery Montague Road Edmonton, London N18	as above	1874	No
Western Cemetery Queen's Elm Parade Fulham Road, Chelsea London SW3	as above	1815 -1885	No

Typeset and printed by Aquilla the inn print shop 4 Gate Street London WC2